ENAMELING

Connie Wagner

Schiffer Publishing Ltd

4880 Lower Valley Road · Atglen, Pennsylvania 19310

Other Schiffer Books By The Author:
Beads & Agate Jewelry To Create Yourself. ISBN: 9780764329982. $14.99

Other Schiffer Books on Related Subjects:
Contemporary Enameling: Art and Technique. Lilyan Bachrach. ISBN: 0764323555. $39.95.

Schiffer Books are available at special discounts for bulk purchases for sales promotions or premiums. Special editions, including personalized covers, corporate imprints, and excerpts can be created in large quantities for special needs. For more information contact the publisher:

Published by Schiffer Publishing Ltd.
4880 Lower Valley Road
Atglen, PA 19310
Phone: (610) 593-1777; Fax: (610) 593-2002
E-mail: Info@schifferbooks.com

For the largest selection of fine reference books on this and related subjects, please visit our web site at **www.schifferbooks.com**
We are always looking for people to write books on new and related subjects. If you have an idea for a book please contact us at the above address.

This book may be purchased from the publisher.
Include $5.00 for shipping.
Please try your bookstore first.
You may write for a free catalog.

In Europe, Schiffer books are distributed by
Bushwood Books
6 Marksbury Ave.
Kew Gardens
Surrey TW9 4JF England
Phone: 44 (0) 20 8392 8585; Fax: 44 (0) 20 8392 9876
E-mail: info@bushwoodbooks.co.uk
Website: www.bushwoodbooks.co.uk

Library of Congress Control Number: 2010921389

Designed by RoS
Type set in Geometric706 Blk Cn BT/Humanist 521 BT

ISBN: 978-0-7643-3444-3
Printed in China

Dedication

I would like to thank my teacher, Doris "Rafka" Silk, for introducing this wonderful medium to me. It has really changed my life.

I would also like to thank my parents, Margaret "Penny" Wright and Thos Wagner for helping me become who I am today.

Last but not least I want to thank my husband and children for being there for me.

Thank You.

Contents

Foreword

My original ambition was to become a writer. I had been an editor for the South Philadelphia High School newspaper. My choice was to major in the Academic course and I selected "Art" as a fifth major. I knew that artists had a tough time earning a living, so I selected Illustrating and Advertising courses, thinking that finding work would be easier in the advertising field. I took a job buying art supplies, crafts, and hobbies for a local department store and found myself enjoying these other artistic, hobby, and craft fields. Trains and flying model airplanes (building them too) were fun, but the craft world was expanding, and I had always done a lot of painting. I especially loved enameling on metal, because color was so important and I was a basic colorist. In addition, opportunities would provide the artistic craftsmen with novel "accidents" to explore.

This book was written by a former student of mine, and the biggest compliment to a teacher is to see a student grow beyond one's own talent.

Doris "Rafia" Silk

Enameling Introduction

I love enameling. It is an unpredictable art. You start with a goal in mind, but you end up with visual surprises because when the glass meets the heat beautiful things happen.

What is enameling? It is the art of melting glass on to metal. You can use gold, silver, copper, or steel, to name a few. In this book, I will only be working with copper. I enjoy working in copper because it's hardy and relatively cheap compared to gold or silver. With enamel you can create "2D" or "3D" art. Some examples are jewelry, bowls, trays, light switch covers, signs, wall art, and decorations for box tops and wind chimes, just to name a few.

Enameling has been used for many years, including but not limited to the Egyptians, who used the process on clay. The ancient Greeks, Russians, and Chinese have used the enameling process on metal.[1] Enameling is around us in our world on street signs, and in bathrooms and kitchens. This is because it is a very durable medium. It will hold up in weather. The word "enamel" is believed to come from the word "*smelzan*," which is High German, and then later from the word "*esmail*," which is Old French.[2]

I have been enameling for several years. I began by meeting an enamelist, Doris "Rafka" Silk, at a craft fair. The way the colors shine, blend, and work together really got me hooked on the art. A few years later, I learned she taught how to enamel. I took many of her classes. Then I got all the supplies for myself and now I do it from my home.

I have read many books on enameling. Most of them have been written by enameling artists with many years of experience, and they have a vast network of knowledge concerning enameling. I wanted to make a book that will help the artist just starting out. This book will show you how to use a few simple techniques and still come out with art you can be proud of. I will show you mistakes that can happen and how to fix them. Also in this book I will show you alternates for some materials. If your piece is going into a museum you may not want to use the substitutes, but if the piece is for yourself or a friend, they will never know the material you used.

One thing you should remember about the enameling techniques in this book is that you must have an open mind as to what comes out of the kiln. Enameling is not like any other medium. You try your best to guess what will come out, but there are surprises. You will see that one of the projects in this book starts out as one thing and finishes as something else. One thing you should do ahead of time to help to prevent surprises is make color test strips, so you can see how your colors react to the metal, because the color in the jar looks different than the color on the metal. Here are some examples. I will be showing you how to create one of these test strips in this book.

I will also be going over basic techniques, just to get the beginning artist started. There are some things I will not be doing. This is because they are difficult to master or they are made with dangerous materials. I want you to be able to try these basic techniques, get satisfactory results, and not get frustrated. After you have mastered these techniques you can combine them and create some beautiful things. This art can be very rewarding. Some of the techniques I will be covering are:

Applying the varied forms of enameled glass to metal:

Preheat the kiln to 1200-1500 degrees prior to each project. It takes about forty-five minutes to one hour.

Powder enamel is made from both opaque and transparent glass. To apply enamel powder, to the metal you use a variety of sized sifters. You can also use enamel powder to make an enamel with paint-like texture, by mixing enamel powder with your holding agent, such as Agar or Klyr-Fire, which you can then paint on to the metal with a paintbrush. Since you make the consistency of the paint yourself, you can make it as thick or as thin as you like, so you can make it like a paste or a watercolor thickness.

Threads are pieces of glass that you can buy in different thicknesses. This is nice because when a thread melts in the kiln it doesn't get much thicker. You can use these in abstract pieces and just let them randomly fall on your piece, or you can use Blu-Stic and place them exactly where you want them to create a more precise design. Blu-Stic is a liquid you can use to bind the glass piece onto the metal almost like a glue. Then when you put the piece in the kiln the Blu-Stic will prevent the glass from moving and it won't affect your design.

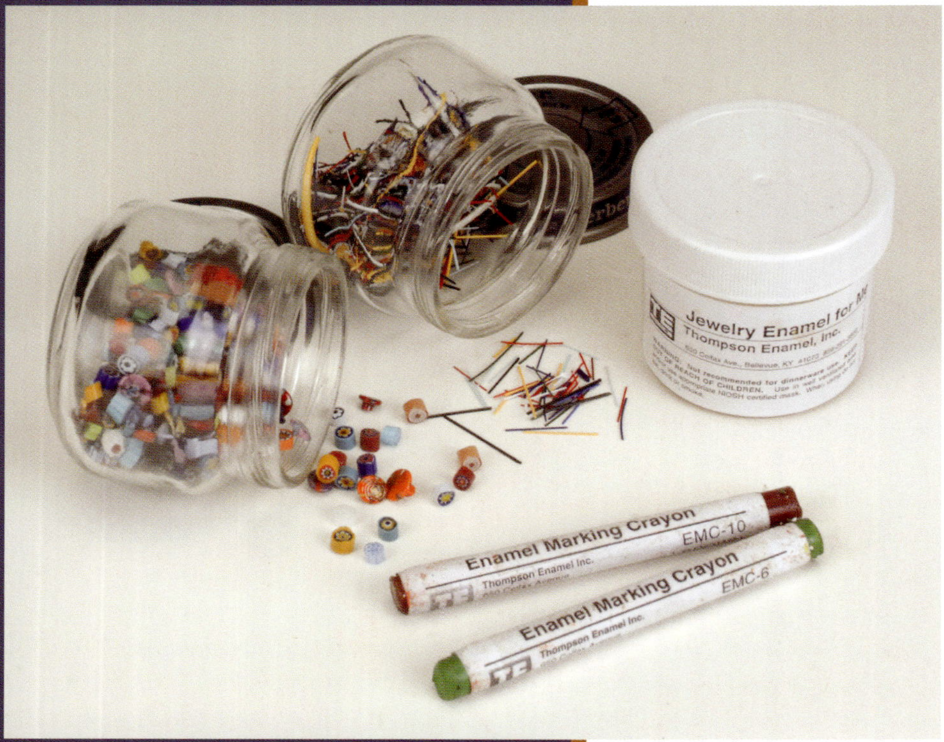

Lumps are pieces of broken glass. You apply them to your piece with tweezers. As with threads, you may place them randomly on your pieces, or use Blu-Stic to make sure they stay where you place them on your piece. Because these are three dimensional pieces of glass, they have a tendency of moving around when you are moving them from your workspace to your kiln. These lumps melt to a variety of sizes, depending on how big the lump is and how long you leave it in the kiln. With lumps you are able to take them out before they are completely flat on the metal, similar to a bas-relief. But you have to be careful, because if you leave them too three-dimensional they could crack off after they cool. Because of this, you must let lumps cool longer than other forms of enamel.

Buttons, wafers, and millefiore: These are all small decorative objects you can add to your piece. They will grow when they fire in the kiln, like lumps. They look similar to flowers. You also apply these with tweezers, and you may use Blu-Stic to make sure these items don't move either while you are moving the piece to the kiln.

I leave pieces made with lumps, millefiore, buttons or wafers on my work space at least 24-48 hours to see if they will crack. Pieces usually cool to the touch in 10-15 minutes. Just remember, even though it is cool to the touch, it may yet crack because it is still cooling internally.

Two techniques for manipulating the glass are scrolling and sgraffito:

Scrolling is a technique that allows you to move the melted glass while it is still in the kiln. Scrolling is the technique for moving the colored liquid glass through other colored liquid glass. The colors won't mix like paint; they will make a design. The results remind me of mixing oil and water. You end up with a swirling effect on the metal. Also during this project I will be showing you how to allow the color of the copper to show.

Another technique is sgraffito; this is the technique of having a fired layer of enamel and then applying a powder layer. Now, before you put it in the kiln you sgraffito the surface, which is scratching a design into the powder layer to show the fired layer underneath. When fired, the design will show. I will be showing this technique with the blades of grass in the flower pin project.

The technique for enameling both sides of a piece is called counter-enamel. This technique is helpful because it prevents the metal from warping and helps keep the glass from cracking off the metal. This is because the glass cools at a different rate than the copper. The larger the piece, the more likely this will happen. Also, with counter-enameling you can make art on both sides of the piece, so you can reverse the piece. I am going do be using this technique on a pair of earrings in this book.

Also other things you can do with glass and non-glass objects on your piece:

A product you can use to enhance your piece is Scalex. You apply this product with a paintbrush. Be very careful with this product because it is poisonous. When you apply this product to your metal, it will prevent any enamel placed on top of it from adhering to the metal. This technique and the others in this book are all unpredictable, as can be seen in one of this book's projects.

One of these non-glass objects is heavy aluminum foil. Over history gold, silver, and copper foil have been used. In this book I will be showing how aluminum works with the enamel. You may want to experiment with other metals, because when put under transparent enamel all the different metals look diverse and beautiful.

Another non-glass item I'm going to show you is layering copper (or the metal of your choice) on top of each other within the piece. You may use silver, gold, or steel if you want, but in this book I'll be using copper.

Most of the enameled pieces in this book are abstract, but I will show you how to make a picture and words using the items listed above to create something more recognizable.

Also, I will show you how to make a stencil for the enamel to enhance your piece.

Additionally, I am talking about junk enamel, which is your trash glass all mixed together, and how you can reuse it. On your workspace, put on piece of paper or paper towel under your work area to catch your excess enamel. When you are done with your layer and your piece is in the kiln, pour the excess enamel in a jar with a lid to save for later. Then you can use this mixture as you use other powder enamel; you can use a sifter or mix it with a medium and paint it on your piece. If you are going to mix it with a medium put the powder in another container with the sifters. This is because you may get other junk in your junk jar like scraps of metal or foil, and you don't want to ruin your piece with a piece of metal sticking off of it. When you put your leftover enamel in a container and use it later, the colors will not blend like paint; they will speckle. So if you have red and blue enamel in a jar it will not appear purple; it will be little red and blue specks of powder, almost like the painting technique of pointillism.

Safety

1~Always wear cotton clothing. If you happen to get set on fire, manmade material burns faster.

2~Wear a mask over your mouth and nose while using the powder. This is powdered glass, and it is very harmful to breathe in.

3~Wear gloves that can withstand high heat because your kiln will be at 1500 degrees. Thompson Enamel Company sells good gloves.

4~Wear heavy-duty gloves while cutting metal if you can. This is because the metal can cut you very easily. If you can't because such gloves are too bulky, be very careful.

5~If you have long hair, put it back.

6~If you are interested in buying a kiln for yourself, I recommend a kiln with a temperature gauge, because then you're sure it is at the temperature you want it. Also, a kiln with a peephole is good because you can check if your piece is done without opening the door; this will prevent the temperature from dropping as much.

7~Once in a while you may have a spark or a fire in the kiln. If this happens, shut the kiln and cut off the electricity.

8~Talk to an electrician and make sure you have an electric line that is strong enough for your kiln. If your kiln takes too much electricity, it may cause a fire in your wall.

9~Position your kiln at least 2-3 feet from every wall.

10~Put your kiln on a fireproof top, such as stone or a metal table.

11~When you take your pieces out of the kiln put them on a fireproof surface until they are cool. I use the Hot Pad/Furnace Floor Pad from Thompson Enamel Company to sit my pieces on to cool. This cooling process usually takes about fifteen minutes.

12~Don't work with a fan in your room because of the risk of blowing the glass powders around.

13~Make sure there is plenty of fresh air. Open a nearby window. Your kiln will get up to 1500 degrees, and the heat will get to you without fresh air.

14~Wear safety glasses while using a drill. You don't want tiny pieces of copper to get in your eyes.

15~Do not have carpets in your workspace. This is because if you are working on carpeting and you drop a piece on the floor you may start a fire, or at the least singe your carpet. Keep metal tweezers close by so if you do drop a piece of hot metal you can pick it up fast and put it on your cooling area.

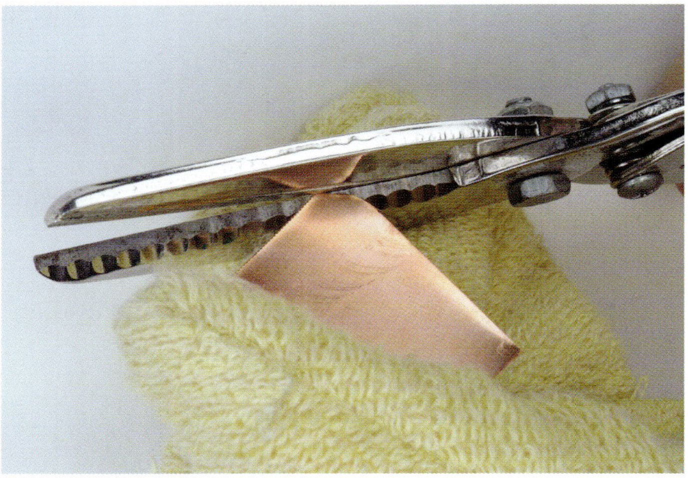

16~Have first aid nearby, because if you get cut by the metal you won't notice it at first, but after a minute or two it will bleed and sting like a bad paper cut. Also, let your pieces cool at least ten to fifteen minutes between firing and working on them again. A piece will look cool before it becomes cool, and if you touch it too soon it will burn you.

17~If you have any open wounds on your hands, cover them with Band-Aids, because you don't want glass powder to get in your wound.

Setting up your workspace and getting ready to work:

1~Make sure you set up the kiln in a safe place (see **Safety**).

2~Make sure you have all your tools at your work area. You don't want to walk away from a kiln that is hot. You may want to bring the phone with you or just ignore it if it rings.

3~Make sure you have a cooling area beside your kiln. When you remove the pieces from the kiln they will be hot for at least ten to fifteen minutes.

4~Work on a floor space that doesn't burn easily (see **Safety**)

5~Put your kiln near your work area. This is because you will be carrying a dust-covered piece and you don't want it to get disturbed in transit.

6~Allow time to preheat your kiln to 1500 degrees. It takes my kiln about one half hour to reach the appropriate temperature.

7~Also, I don't recommend working for more than a few hours near the heated kiln. When I work, I get so involved in my work I don't realize how bad the heat is effecting me. It gets very hot. You may want to have a bottle of water with you.

You need a variety of heavy gauge metal. You can use gold, silver, steel, or copper to name a few. I prefer 18-gauge copper. It is more difficult to cut but it doesn't warp as much in the kiln as compared to thinner copper. You can buy it precut into shapes (I usually find my precut shapes online from Thompson Enamel—or if you are lucky on eBay) or buy it by the sheet (from a hardware or roofing supply store) and cut it yourself.

Tools that you should have but are not limited to:

Some other things that I recommend you should have that are not in these photos are an old-fashioned iron and first aid. Depending on the project you are working on, you may want jewelry findings, like jump rings, clasps, hooks, earrings, or hardware pieces to hang the project on the wall, pieces of wood to mount the project to or things to adhere your piece to, such as a box or frame.

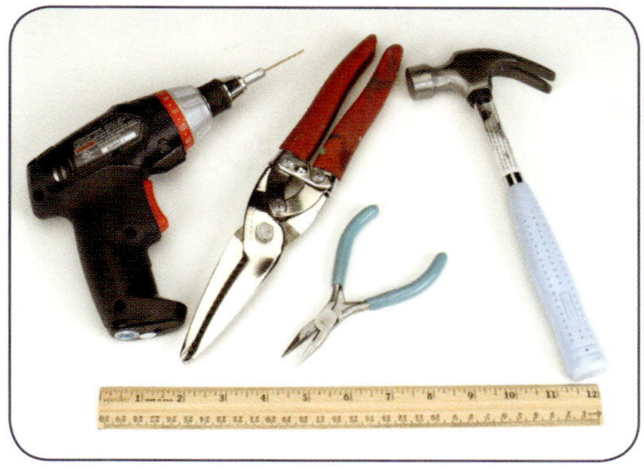

The hardware tools are a drill with a 1/8-inch bit, heavy-duty cutters, pliers, a hammer, and a ruler.

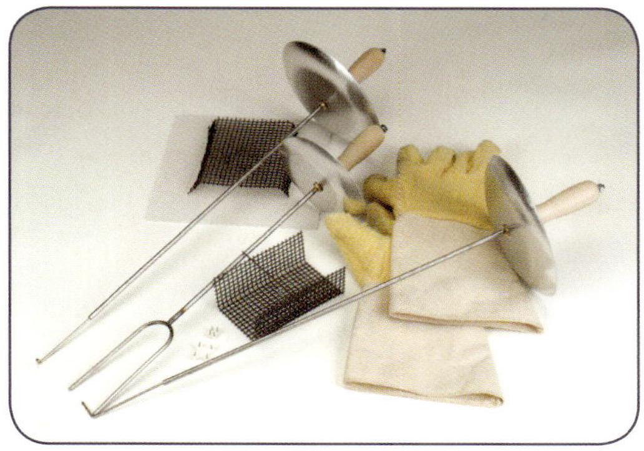

These are the firing tools, fire proof gloves, bead rack, firing rack, fire proof pad, to put in your kiln or use as a cooling area, and scrolling tools.

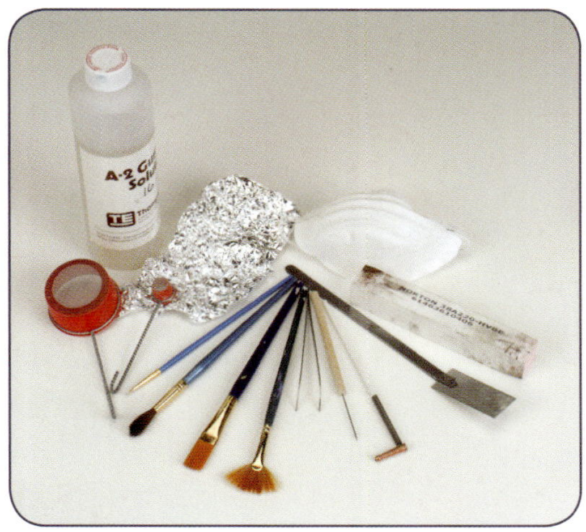

These are some application tools, as well as other things, including Gum Solution, three different sizes of sifters, a variety of paintbrushes, alundum stone, to file, tweezers, a scratching tool, aluminum foil, and a dust mask.

Some different mediums you can use are Klyr-Fire or Gum Agar

Tips

1 ~ You can use hair gel instead of Gum Agar and Klyr-fire.

2 ~ You can use heavy aluminum foil instead of silver, gold, and copper sheet foil.

3 ~ If you put enamel powder in an area on your piece and you change your mind, you may remove it before you fire the piece with a clean dry paint brush.

4 ~ If you are making a smaller piece (usually under 2 x 2 inches), you don't need to counter-enamel it. To finish the back of your piece, painting it with nail polish (in a color to compliment your piece) works well, as demonstrated on the bracelet project.

5 ~ Cover your area where you apply the enamel with paper, newspaper, etc., but have your piece on a smaller piece of paper, or paper towel. You will need to be able to pick up the smaller piece of paper to dump the remnants of the enamel in your junk enamel container.

6 ~ If you pre-drilled your piece, before you put it in the kiln, make sure the holes are not covered with enamel. If you don't, the enamel will clog the holes, and if you have to drill the holes again there is a good chance you will break your piece.

Cleaning

Cleaning your copper:

You must clean your metal before you work on it. This is because metal picks up a lot of oils and dirt from your hands and the environment. To clean your metal you must use a mixture of a gritty cleaner like Oxyclean or Comet. Also you need liquid dish soap. You mix both cleaners on a scrubbing sponge. You then scrub the metal on both sides while holding the piece by the edges. Then dry the pieces with paper towels. I usually leave the pieces in the paper towels and carry them back to the workspace and leave them in the paper towel until I'm going to use them. Then I pick each piece up by the edges and lay it on my workspace. If you don't clean your metal, the oil from your hands will leave areas that your medium will not stick to, so your enamel will come out of the kiln spotty and burnt, as shown in this photo.

Cleaning your workspace:

You should clean your workspace after every use. This helps because you won't waste enamel; you can put it in your junk container and use it another time. Also, you don't want old color dust on your workspace because it could get on the next piece and ruin it. After I dump the leftovers in the enamel container, I use a general cleaner and loads of paper towels to wipe up what's left. The reason I say loads of paper towels is because it is glass you're cleaning up and it can still cut you, especially the lumps and threads, as will the copper.

Also make sure to clean up and dispose of the cut off pieces, as they are very sharp and dangerous. They will cut right through the sole of a shoe.

Above everything else, just enjoy enameling; it is very rewarding.

Endnotes

1 http://www.ive.org.uk/faq/history.html
2 http://www.ive.org.uk/faq/what-is-enamel.html

1

Test Enamel Tile

For the first project, I will be making a test strip. Because the enamel looks different before and after it is fired it is good to do these so you will know how the colors will look. I have applied the enamel four different ways so I can see how the enamel reacts with the metal. In the top quarter I have placed one coat of the enamel on the copper backing. The next quarter has two layers of enamel. The third quarter down has the enamel on white enamel. The fourth and bottom quarter is a layer of enamel on foil, which adds a different effect, to the enamel. Each technique provides different color effects.

This photo provides an example of four different colored test strips: top left is yellow; top right, black; bottom left, robin egg blue; and bottom right, green.

I start with a piece of copper I cut myself and then cleaned. Next, use a paintbrush to apply Gum Agar to the copper.

I'm working with Harold Purple enamel. One quarter will have one coat on copper, the next quarter of the copper will have two coats of enamel, the third quarter will be one coat of purple on opaque white, and the forth quarter will be one coat on aluminum foil. I'm using a piece of paper to cover half of the test strip while applying the first coat of purple to half the strip. I'm using a sifter to apply the powdered purple. It is a 1/2 x 1/2" sifter with a 40 mesh screen.

The copper strip is now half covered with the Harold Purple enamel powder in a single coat. Put more enamel on the edges because it will burn out first there.

Cover the area you previously covered with Harold Purple with a piece of paper, cover the last quarter where we will be applying purple enamel on aluminum, and put down a layer of white enamel for our purple on white strip.

Apply a layer of powdered flux (which is clear) to the last quarter. We need this as a base to lay down the aluminum on after the first firing.

The initial coating of Harold Purple, white enamels, and powdered clear flux have been applied. Here's what it looks like prior to firing. As you can see, the colors look very similar in the powder form. That's why I suggest making these test strips.

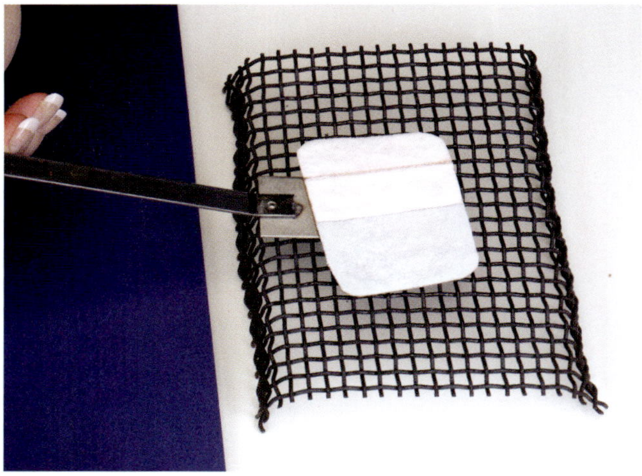

Using a small spatula to carefully move the coated strip from the paper to a firing rack prior to placing it in the kiln for the first firing. Make sure you have the strip ready to go when you open the kiln door, as the temperature will drop a bit just from opening the door.

The test strip is straight out of the kiln after the first firing. A firing takes from one to two minutes. When the enamel looks like an orange skin, smooth with small bumps, it is ready to be removed. If it still looks like powder, it is too soon. If it looks black, it's much too late. Warm colors and fluxes heat faster than darker colors. Check the warmer colors after they have been in your kiln for 1 1/2 minutes. They may take as long as 3 minutes, but not usually. Check the cooler colors after 2 - 2 1/2 minutes, although they could take as long as 4 minutes. The timing for the colors are true if the kiln is at 1500 degrees. If it is cooler, then you need to leave the pieces in the kiln longer.

I am now ready for the next step.

Paint a fresh coat of Gum Agar over the entire piece to make the subsequent layers of enamel stick.

I am now applying Flux over the one coat of purple. This prevents the enamel from burning when you fire it again. Cover the rest of the piece as before with paper to prevent the enamel spreading to other areas.

Apply Blu-Stic and the aluminum foil to the last strip. The Blu-Stic will hold the aluminum in place and burn off in the kiln.

Apply a coat of Harold Purple over the other three quarters of the test strip, covering the first quarter with paper to retain its single coat color.

The test strip is ready to return to the kiln.

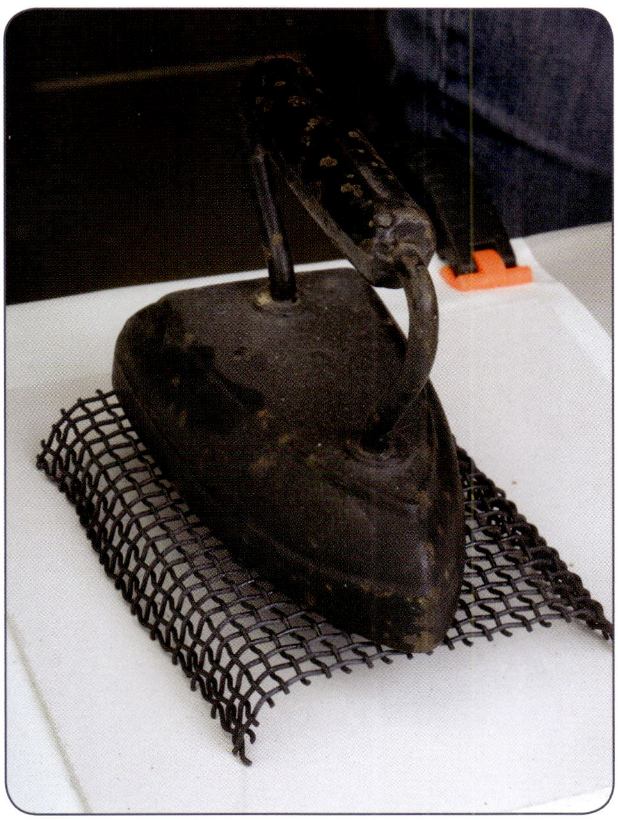

I place an antique flat iron over top of the strip just out of the kiln now to keep the aluminum flat and to keep the piece from warping.

The test strip is done. We have one quarter with one coat of Harold Purple, the second strip with two coats, the third strip is Harold Purple on white, and the final quarter is Harold Purple over aluminum. This shows you how the coat of color will look and behave in different applications.

2

Reversible Earrings

Now I am going to make reversible earrings. This is done with the technique counter-enameling.

Now we're making earrings. Note that this is precut and smoothed copper, so the edges are not sharp. Drill holes in both earrings about 1/8" from the outer edges. One thing to note when making earrings is that you need to decorate and fire them at the same time or they will not look like a set. The variations would just be too extreme.

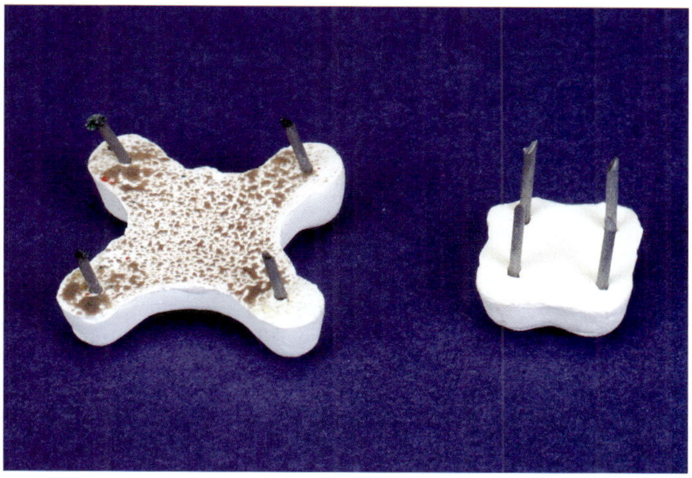

We are going to coat both sides of the earrings with enamel, so we need stilts to keep the hot enamel from sticking to the stand in the kiln. With the stilts, the hot glaze only sticks to a few small prongs and is easily removed.

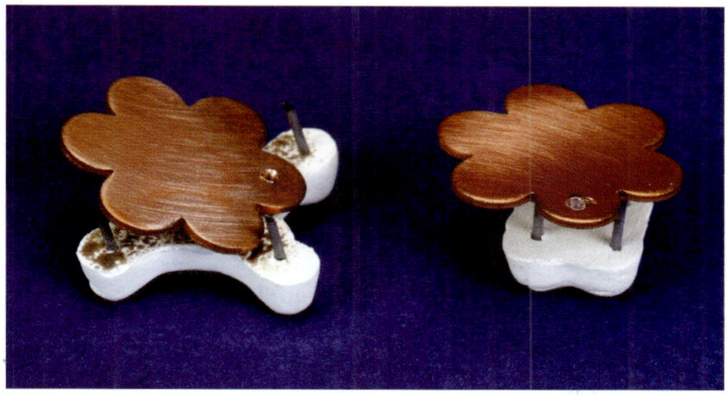

Make sure the stilt is the right size for the earring, as seen with the right hand piece. If it's too large, the stilt cannot work properly, as seen on the left sample. Do this before you start applying your enamel; otherwise you will mess up your design.

First cover the surface of the earring with Gum Agar using a small paintbrush.

Applying Opaque Butter Yellow enamel to the agar coated earrings.

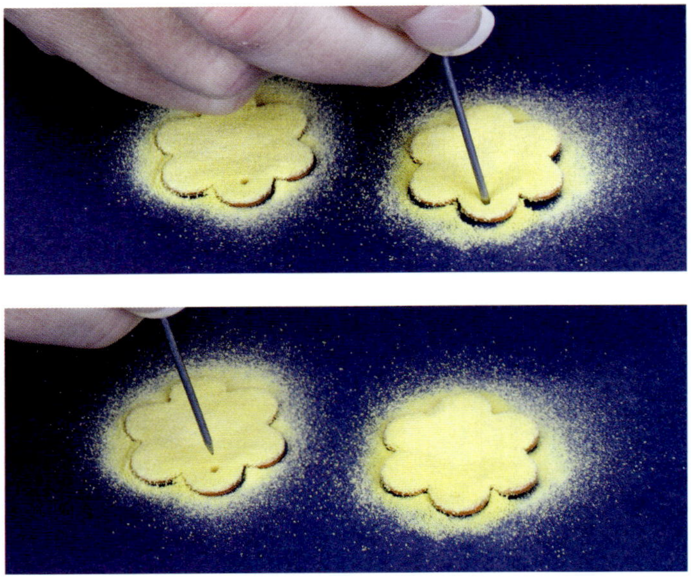

I'm using my sgraffito tool to clear out the holes in the earrings. If you don't, the fired enamel will fill in the holes, leaving you with teeny tiny pendants.

Now I'm adding small amounts of "junk" to the back of the earrings to make them more visually appealing. "Junk" is all the leftover enamel that falls on your paper while you are coating the copper in enamel. Save all the different scattered enamel powders in a "junk jar" to give yourself a rainbow of powdered enamels to use as visual accents at times like these.

Use the spatula and tweezers to move each earring to your firing rack. This is a challenging step when working with small objects. Then you put them in the kiln for 1-2 minutes.

Hot out of the kiln. Allow the earrings to cool before the next step. This will usually take 10-15 minutes.

Turn the cooled earrings over and apply agar to what will become the front sides.

Adding red enamel to the agar now.

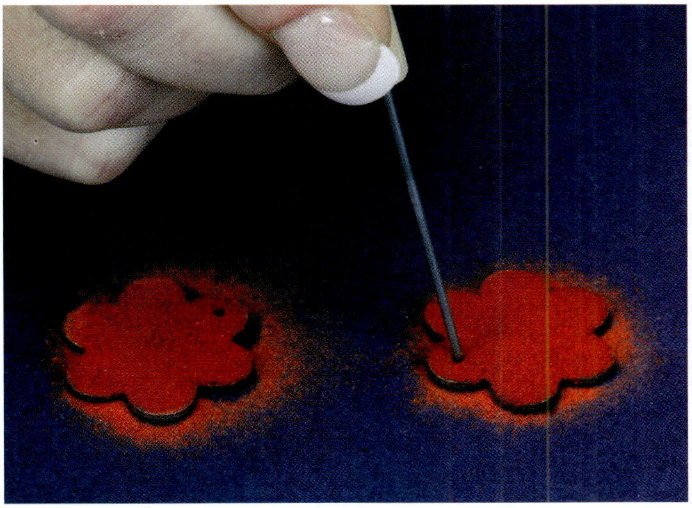

Don't forget to clear the holes of enamel now.

Place the earrings carefully on the stilts now, as you don't want the enameled backs to stick to the firing rack during their reheating. It is easy to remove the earrings from the tiny points of the stilts; it is not easy to remove them from the rack if they stick to that.

Once the red has been fired, the stilts need to be removed (the softening and cooling of the enamel on the base has stuck the stilts to the earring backs). Use your spatula and tweezers to accomplish this.

One reason for applying an interesting design to the back of the earrings is that the design hides the stilt marks.

I'm adding Transparent Lime Yellow around the edges of the earrings for interest. Note: one of your big challenges is going to be keeping the earrings balanced on the stilts and preventing them from falling off as you move them into the kiln. It may take several tries to get them into the kiln with the earrings balanced on the stilts. If the earrings should fall off the stilts as you place them in the kiln, leave them sitting aside for several minutes to cool before you try to get them sitting atop the stilts again. Remember, the interior of the kiln is over 1,000 degrees!

A word to the wise: if you try to remove the earrings from the stilts too soon, while the enamel is still hot, you end up with painful spikes of enamel that just won't do. Reintroducing this earring to the kiln for 20 to 30 seconds will melt down this spike. When you do this, put both earrings into the kiln together again so they will look like a matched set. Always remember, though, if you have enamel on the back of a piece, always put the pieces on the stilts.

I've decided to experiment with the earrings. I've added Blu-Stic and a lump of broken yellow glass to each earring, a coat of agar, and then clear glaze around the edges so they won't burn. I'll heat it all and see what we get.

The end result of the earring firing isn't bad. More experimenting could be done.

Adding rings and fittings to the earrings with needle nosed pliers.

The finished earrings. Note that the earrings are reversible. This doubles the use of each pair of earrings.

3

Bracelet

The next project is a multi-piece bracelet. I will be showing how to use foil and lumps within a project, as well as an easy way to finish the backs.

Drill holes in either end of each copper piece. Since the copper is curved, you will drill a straight hole if you hold the copper at an angle. You will also have a firmer surface to drill through.

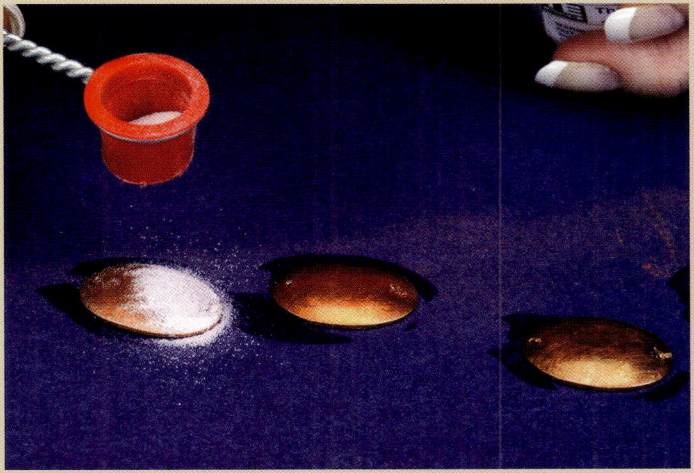

Add Gum Agar and purple enamel to each bracelet section.

Like so. Now fire them in the kiln.

Hot out of the kiln. Now, since these pieces are all convex, some of the edges have lost their enamel. I'm putting on another layer of enamel to coat the edges properly. If you want your pieces to have these edges without enamel, coat them with flux before the next firing so they don't burn out.

Applying the second coats so the bracelet pieces will be completely covered.

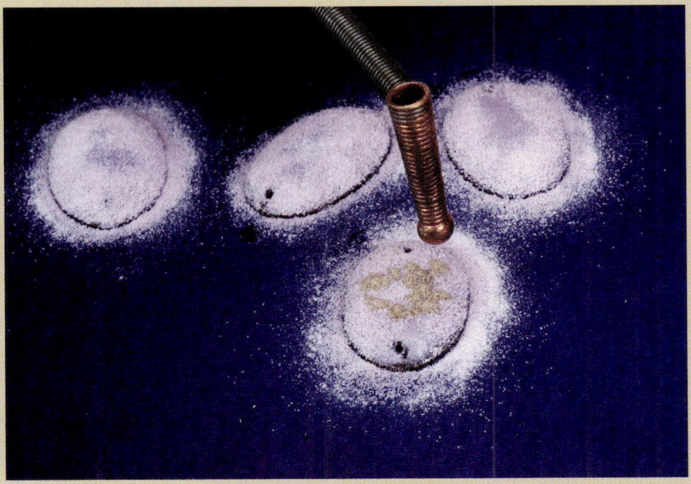

Applying Transparent Van Dyke Brown with a small sifter.

The bracelet pieces that received the Transparent Van Dyke Brown came through the firing well and have an interesting look.

Adding Blu-Stic and aluminum to the two pieces of the bracelet that will be decorated with aluminum foil and enamel.

Coating all four pieces with Gum Agar.

Adding Transparent Van Dyke Brown over the foil covered pieces.

Adding a bit more Transparent Van Dyke Brown to the other two as well.

Adding lumps of crushed glass to the two pieces without the foil. At this point, you can control how you want these pieces with crushed glass to look. If you leave them in the kiln for a short time, the lumps of glass will maintain their shape and give these pieces some extra three-dimensionality. If you leave them in longer the lumps will melt flat. Be careful when you use lumps and al ow a few days for the piece to cool, because since the lumps cool at a different rate than the other glass and the metal, they are apt to fall off. So don't finish your piece with jump rings, clamps, etc. for a few days.

Notice how the glass lumps on the bracelet pieces have melted down entirely. The bracelet enameling is complete.

If the back of the bracelet is never going to show, once the firing is done paint the backs with nail polish to give them a finished look.

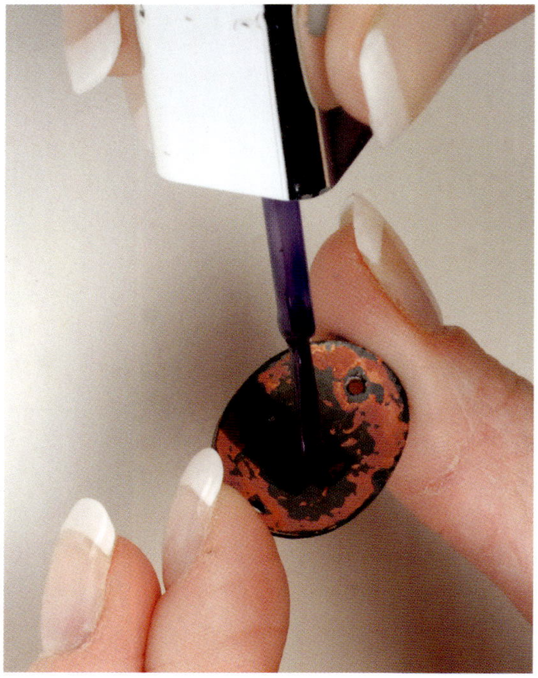

Like so.

Use an Alundum Stone file block to file down the edges of the enameled bracelet to get the copper edge to show. You must do your filing under water to keep the enamel from cracking.

Now I'm going to be doing the finishing touches. Using 18 gauge colored copper wire and glass beads, make the bracelet chain. I create and close the loops at either end of the chain with needle-nosed pliers.

The bracelet has been assembled.

4

Multiple-pieced Pendent

For the next project I am going to be making a multiple-pieced pendant. During this project I will show you how to layer copper, make and use stencils, use enamel threads, and use Scalex.

First I will be drilling holes in the copper.

I'm applying Scalex to areas where I want the copper to show through. Scalex will prevent enamel from sticking to these areas. After the piece comes out of the kiln, I should be able to clean the area and remove the Scalex, as well as any enamel on top of it.

I'm using a thinner gauge copper scrap to add an extra dimension to the piece. Keep a box of scrap copper for times like this. Waste not, want not, as they say.

Apply Gum Agar and Alpine Green enamel to each piece.

Adding Opaque Butter Yellow to the scrap to be added for contrast.

Adding Transparent Brown to the three pendant pieces randomly for interest.

I've decided to add a little light blue to lighten up the look a bit.

The necklace pieces have been fired now. With a little rubbing with a paper towel, the area that received the Scalex will be glaze free.

Enameling is unpredictable. Sometimes Scalex applications only provide spotty results, as you see here. Don't come to enameling with a preconceived notion of what the end result should be or you will be forever dissatisfied.

Add a little Blu-Stic to hold down the yellow glazed copper scrap. Now we'll move on to a little stenciling for our necklace.

Apply agar to all of the necklace pieces.

You can use a paper stencil to create patterns in glaze. In this case I've cut ovals in paper to create this pattern in orange. I think I'll add some Robin's Egg Blue as well.

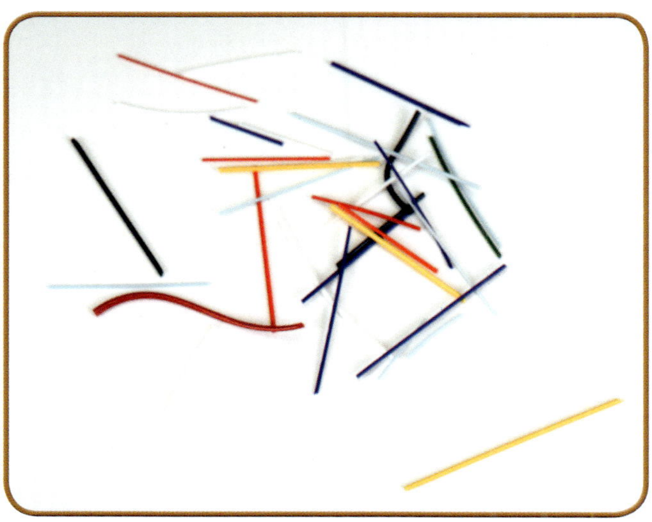

Here are some glass threads. They add interesting patterns to the piece as well.

The threads are in place.

Cover the entire piece with Flux before firing them in the kiln. This clear Flux will protect the exposed copper and help bind all the other pieces together.

The necklace pieces fired well.

Add Gum Agar to the backs of the necklace. I'm following that with a light, transparent blue on the smaller pieces and the dark edges of the larger piece. I like the appearance of the center of the large piece as it is.

I'm adding Flux to the middle of the large piece to retain the colors present. Getting ready to fire it in the kiln. Since I counter-enameled the piece I need to get the stilts ready.

As you can see, the piece fell on the stilts in the kiln. Sometimes it requires something other than a gentle touch to get enameled pieces off of the stilts.

I've decided to add color to the back of the central piece of the necklace. First add Gum Agar as always. Then I add Opaque Robin's Egg Blue for contrast.

Add flux to coat the back of the piece and fire

The necklace backs are finished. This necklace is not intended to be reversible, but this gives it a more finished look.

There are times when your holes do fill in. Here I'm drilling the holes open again, which is a challenge with enamel. You have to be careful not to crack the enamel. I've been lucky so far. You have to make a decision, take the chance of breaking your piece, make it into something else, or put it in your junk box and use it on another piece.

The enamel work is done on the necklace. Now I am going to create a chain out of 18-gauge wire and beads.

Repeat the stringing process with the necklace using beads and 18-gauge wire. Be creative.

Here is the finished multiple-pieced necklace.

5

The Pin of Flowers

I am now going to show you how to create a pin with flowers on it. I will be showing you the techniques of stenciling, enamel threads, enamel buttons, enamel wafers, enamel millefiori, and sgraffito.

We are going to start with a cleaned copper circle.

Add agar to the surface of the disk.

As this is going to be a scene of flowers in a field, cover the bottom half with paper and coat the top half with Opaque Robin's Egg Blue for sky.

On the bottom half use Opaque Green for a grassy field.

Now I am going to fire it, then continue putting more detail in it.

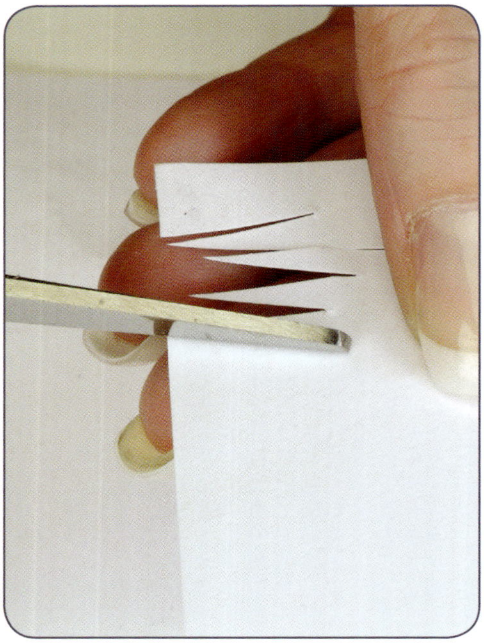

Now I'm going to put more detail in the grass. Cut out a paper stencil in the shape of grass blades.

Before you apply the agar, place your stencil on your piece to make sure you are happy with the design.

Now that I am ready with the stencil I have removed the stencil to apply the agar.

Now apply Transparent Green using the stencil to create grass. I am using transparent green on opaque green to add texture and depth to the grass.

Some of the lines weren't as sharp as I'd like, so I'm adding details with this tool.

Like so.

I'm adding Transparent Blue to the sky in organic patterns to give it depth and texture also.

Add the transparent flux over everything to coat it. You should do this because if you don't there is a chance the areas where there is one layer of enamel could burn out, especially around the edges.

Transfer the pendant to the heating rack and place it in the kiln.

Progress on the pendant so far.

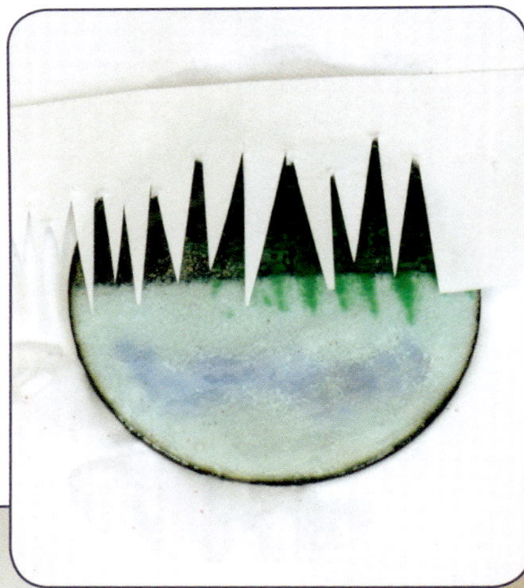

I am going to do another layer of enamel with the stencil to make the piece more interesting. Use some agar to hold a grass stencil in place.

Applying Alpine Green over the stencil.

Remove the stencil and define the blades of grass with your scratching tool.

When green gets into the sky brush it away with a soft, feathered brush.

Now I am going to place the stems. First, I'll apply a coat of agar to help the stems not move. Then I will place the enamel threads, using tweezers, to represent the stems of the flowers.

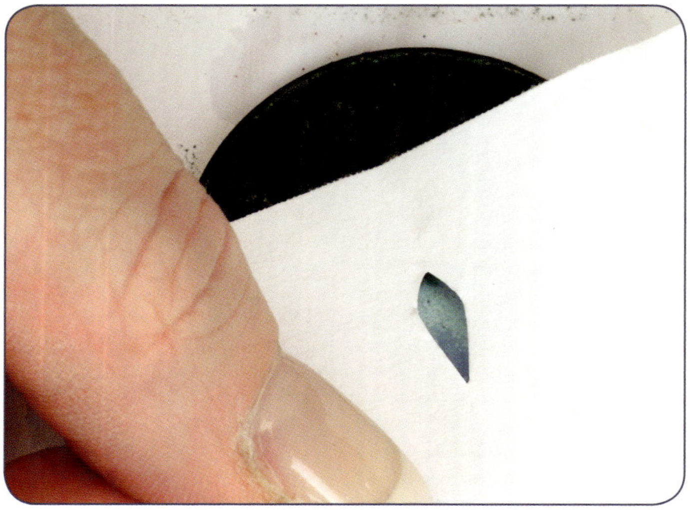

Cut a leaf stencil to add leaves to the flowers on the pendant.

Add leaves to the flowers with your stencil. Just a few will do.

I added a millefiori in the center, and the one on the right is a button for the flower head.

The flower on the left is a wafer. Then I fired it. With buttons, wafers, and millefiori, wait at least 24 hours before you do the finishing touches, such as pin backs, because when these cool they could fall off.

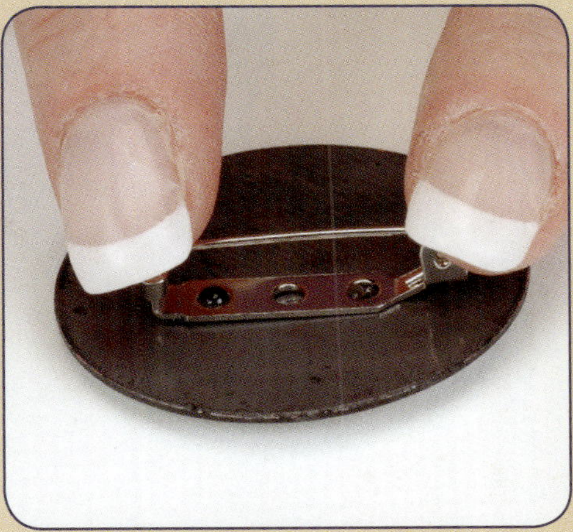

I use 527 Glue by Beacon to fasten pin backs to my pendants. When you attach a pin back vertically, make sure the hinge is to the top so that if it comes open the pin will not fall off. When fastening the pin back horizontally, make sure the opening is to the top for the same reason.

The finished pin.

6

Enameled Bead

In this project I will show you how to use enamel crayons and how to make and use enamel paint.

Here is the bead rack I'll be using to decorate the bead.

When dusting the bead with enamel, put a paper towel between the bead and the firing rack so no enamel gets on the rack. If it did, the enamel would melt in the kiln and stick the rack to the kiln.

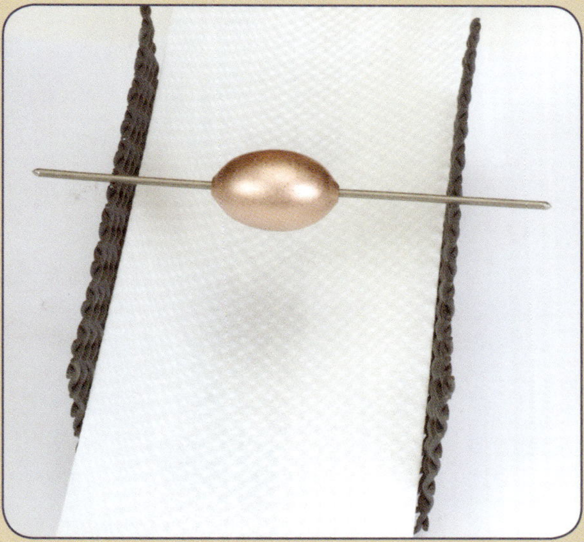

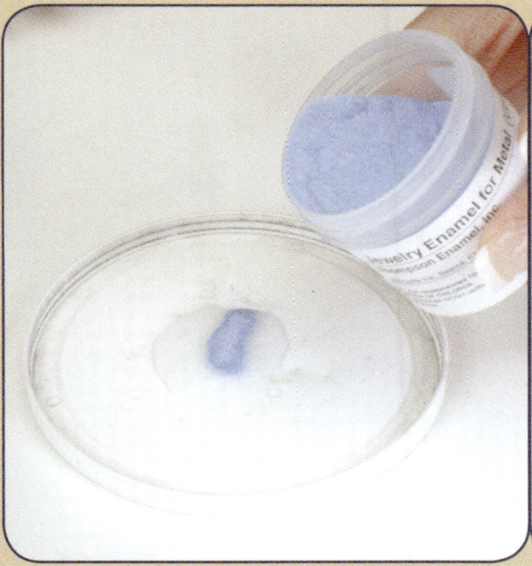

Instead of applying the agar, and then the enamel to the copper, I am going to mix the two together. This technique helps keep the enamel on the three dimensional shape. This technique can be used on any piece of copper and, depending on how you proportion the agar and enamel, you can make a thin or thick paint-like texture. For the first application on the bead I am making a thicker mixture. If you make a thin mixture, first it could burn out.

Apply the enamel agar mix with a small brush. Don't apply enamel too close to the bead holes where the bead is threaded through the rod; the enamel will spread to cover those areas. If you get the enamel too close to the edges, the enamel will cover the rod and the bead will have to be removed by force with a hammer.

Like so.

6

The agar and enamel mix is applied. Now remember, this is a frustrating process, as gravity is working against you and beads spin as you apply the enamel.

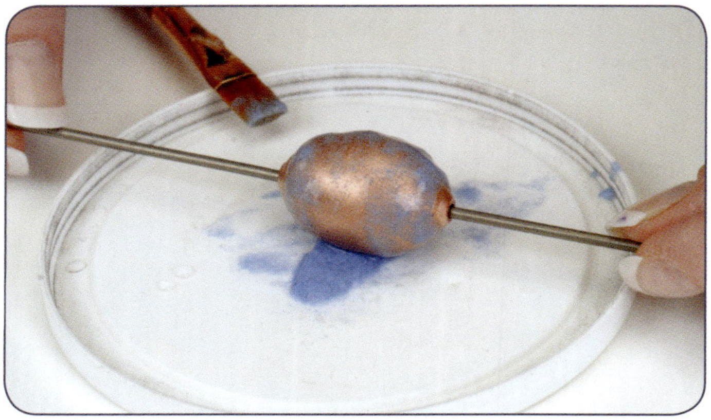

If you feel the brush isn't getting a thick enough enamel coat on the bead, you can roll the bead through the enamel mix on the shallow lid you mixed the enamel and agar in. As you can see, the bead is not completely covered. During this project you have to repeat these steps many times to completely cover the bead and give it an interesting look.

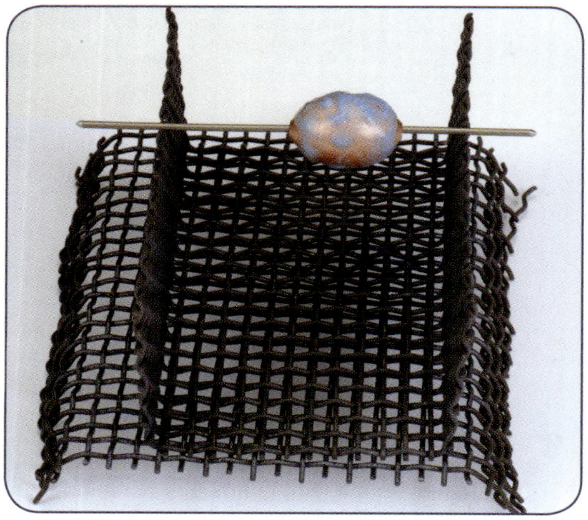

Place the bead rack on a firing rack so you have something you can pick up with the tongs once you slide it in the kiln.

The first firing of the bead: Multiple firings are required to completely cover the bead. The dark areas are where burned copper is showing through.

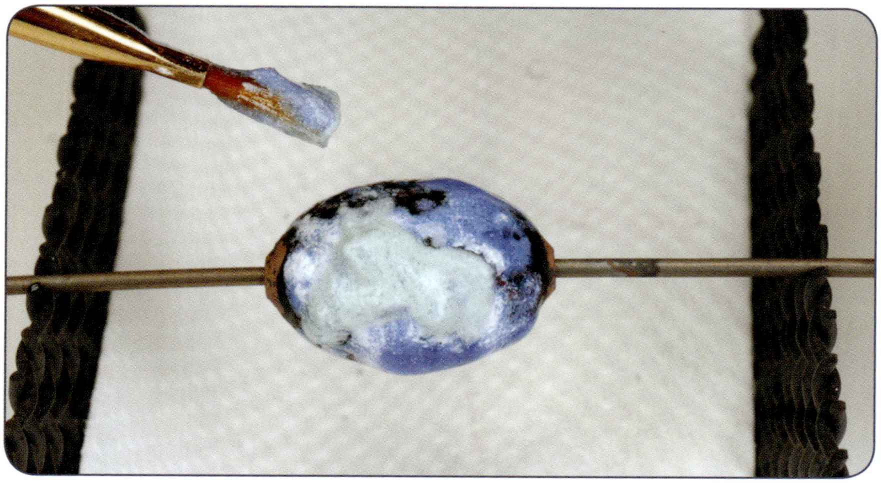

I've applied several layers in separate firings to the bead. The colors applied were Transparent Bonnet Blue, Opaque White, and Opaque Daphne Blue. Now, I'm adding Transparent Grass Green and Opaque Robin's Egg Blue. Note how thick you have to apply the enamel to the bead to counteract the force of gravity. Between each firing you must allow the firing rack to completely cool, because every time you apply enamel you must put a paper towel under it to catch the enamel. It usually takes ten to fifteen minutes to cool each time.

Now I'm having fun applying globs of Transparent Grass Green. This will now be my fourth firing. You have to be patient with enamel beads!

The bead firing turned out well. I accidentally dropped some enamel on the firing rack.

When I am done my project and the rack is cooler, I will use a hammer to get it off.

Now we'll apply a design in Enamel Marking Crayon. First, however, we have to create a smooth surface. For this we fill in all the lumps and valleys with Flux. Make sure you allow each coating to dry before firing. This only takes a few minutes.

Apply the crayon in any pattern you wish. I am using the tweezers to keep the bead still while I am designing it.

Adding flux after the crayon is applied.

I am done enameling the bead. Now you can use your imagination to use the bead in a piece.

7

Writing Words with Enamel

During this project I'm going to make a necklace centerpiece with a word written on it in glass, along with layers of metal for decoration.

I'm using 18-gauge copper.

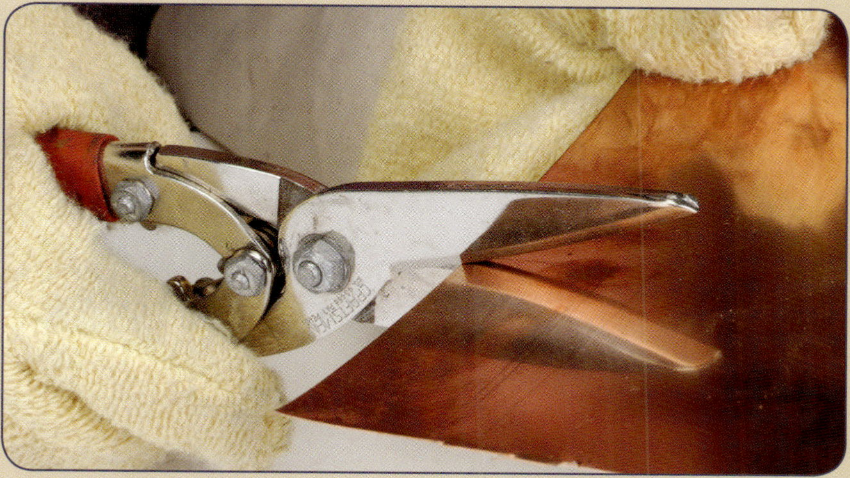

Using metal sheers to cut the copper. Wear your protective gloves or the sharp copper edges will cut you.

I'm cutting off the corners, as they are very sharp. If the edges are still too sharp, after the enamel work is done you can use your file and water to smooth those edges down.

Drill holes in the copper for the necklace fitting. Still wear your gloves, because if you lose your grip the metal will cut you while you are drilling.

I like to hammer the sharp and slightly curled edges of the cut copper plate flat before enameling.

Wash your copper and apply a layer of agar as before.

Add Opaque Cork Brown enamel to the agar. Remember to put more enamel along the edges so they will not burn out.

The layer of Opaque Cork Brown enamel is applied.

Applying a line of Opaque Mocha Brown around the outer edge. As you can see, you can apply multiple layers of enamel before you fire.

Applying Transparent Van Dyke Brown over the line of Mocha Brown.

Adding a line of Transparent Hazel. The center is being left open for the lettering.

Applying a line of Transparent Orange Red Ruby.

Don't forget to open the holes with your scratching tool or a scrap piece of wire.

The necklace nameplate has been fired. The applications of opaque and transparent enamel are looking good.

Now I am going to put enamel on the copper circle for the "O" in love. Don't forget to wash your copper first. Then I am going to apply agar.

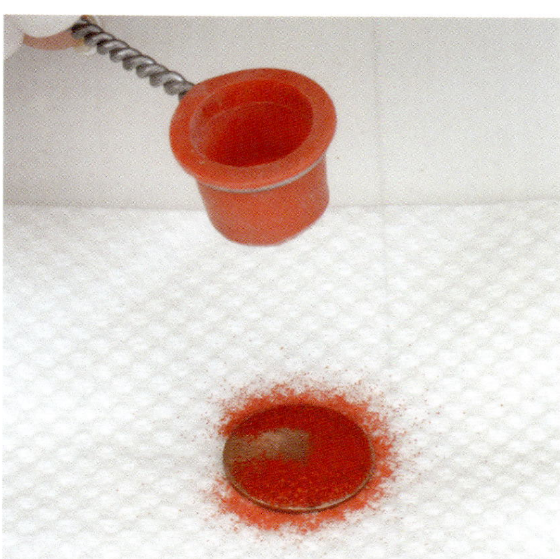

Add Opaque Flame Red to the agar. I'm firing the red "O" separately, as red fires very quickly—in only about 30 seconds to a minute.

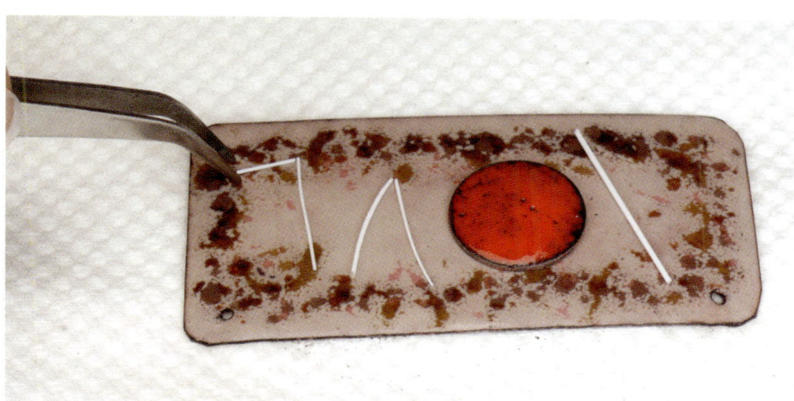

Add agar over the nameplate and add the now red disk and glass rods to form your word, "love" in this case. Use the tweezers to place the threads where you would like them. If the threads are the wrong size break a bigger one instead of putting two together. If you put two pieces together, there is a chance they will move when you put the piece in the kiln and it will change your design.

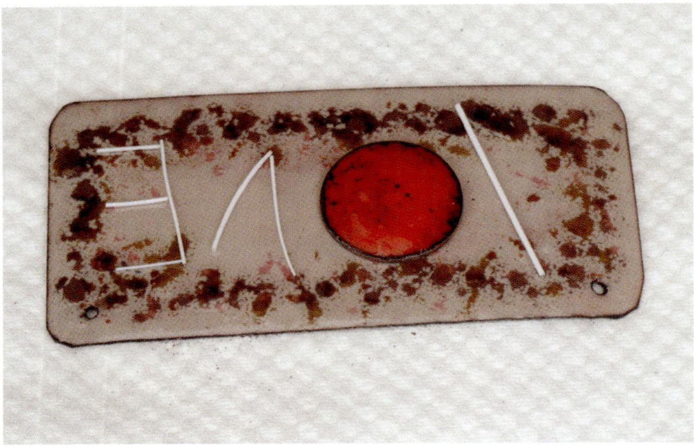

All of the pieces are in place.

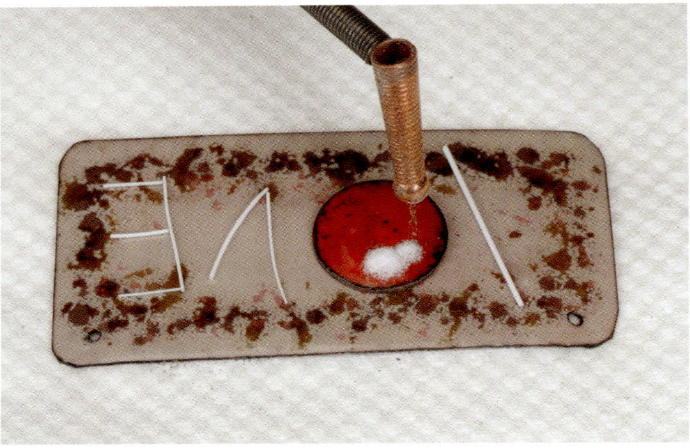

Now, I'm adding Opaque Crackle White to the red. This layer will crack in interesting patterns to enhance the look.

I've decided to add a bit more Transparent Orange before firing.

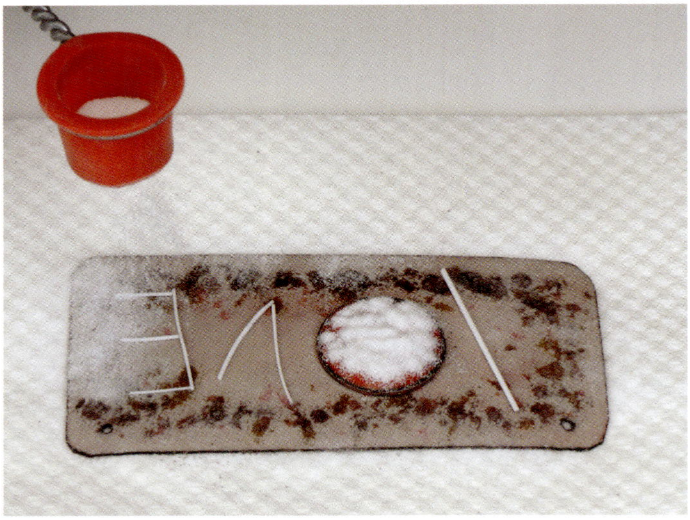

Lastly, add flux over the whole piece before firing it in the kiln. Put a significant amount of flux along the edges where the two pieces of metal meet to keep them together.

Things don't always turn out as expected with enameling. The crackle enamel didn't crackle and the letters are too light. I'll go over the letters with another color to make them stand out. I coated the area of the letters with agar so the colored letters would remain in place.

The green threads really improved the look of the "Love" pendant.

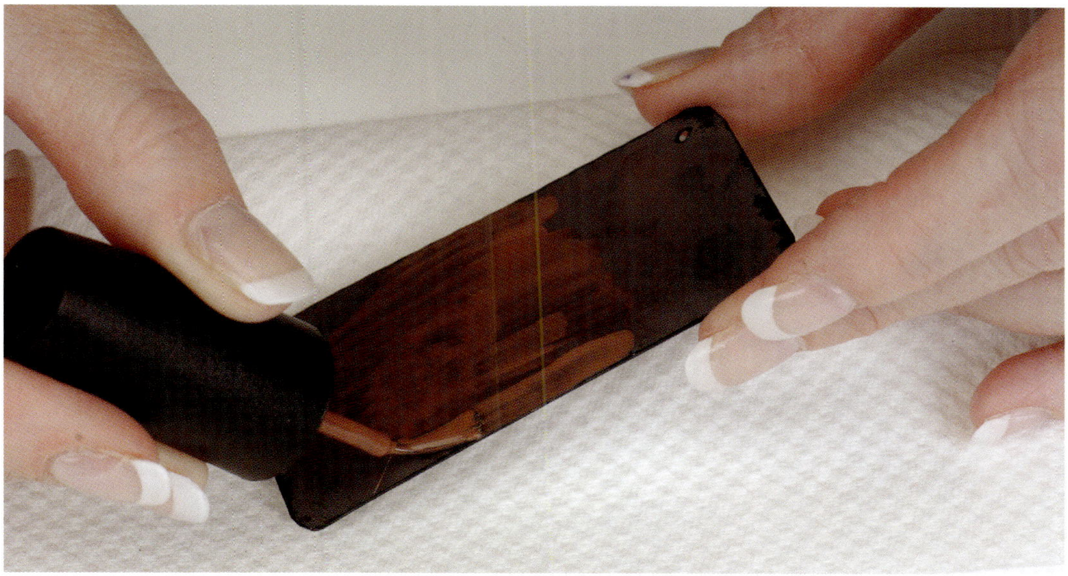

Adding a layer of nail polish to the back of the pendant just to give it a more finished look. Don't use this for a piece you intend to sell as it is not a high end look. Also, when applying nail polish, make sure you wait a few days because pieces cool internally for a few days and may still crack. After you apply the nail polish, you will not be able to put the piece back in the kiln.

I am now finished the enameling part of the necklace.

Here is the finished necklace.

8

Scrolled Pieces

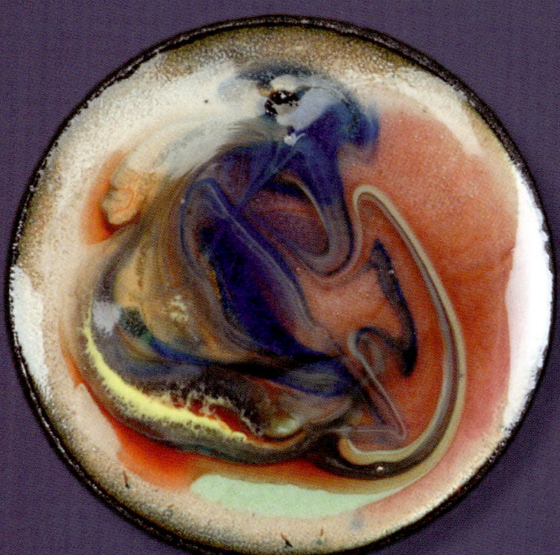

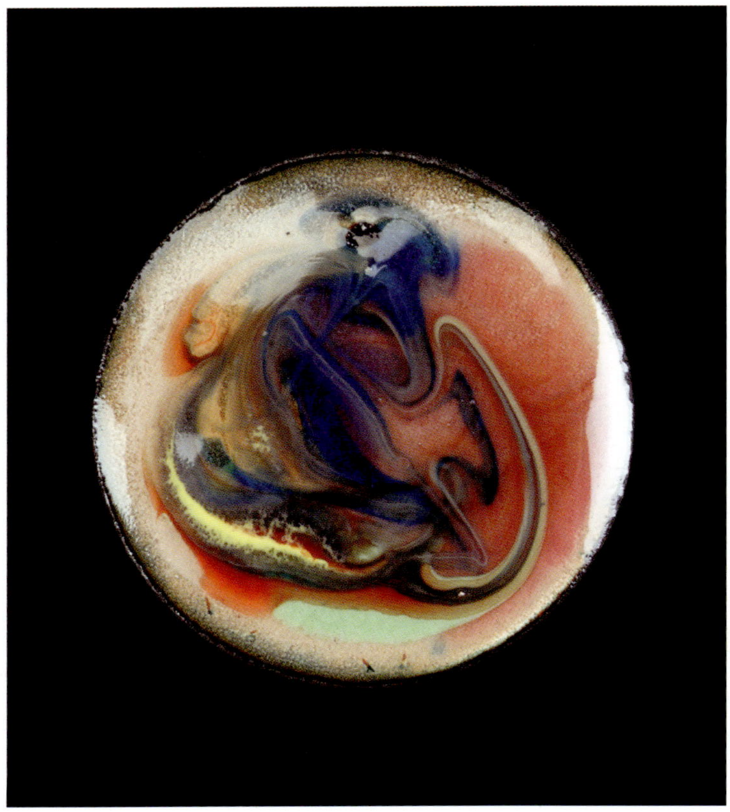

Now I am going to make a piece that began as a pendant but will become a pin. As I have said, enameling is unpredictable. In this piece I will be scrolling the enamel lumps on the copper.

I am drilling a hole in the piece.

Make sure you clean your copper after you drill it to remove all the oil from your hands. While you wash your copper check it for a layer of plastic. Some copper comes with a thin layer of plastic backing. You must make sure to remove this plastic before firing the copper in the kiln, otherwise the results will be nasty and you'll most likely have a fire on your hands.

Add Gum Agar and flux to the pendant. I am putting flux straight on the pendant because I want to keep the copper color in the background.

Add lumps of glass to melt in the kiln. Don't put lumps near the edge of the piece because the lumps could melt over the side of the piece.

Make sure your kiln is 1400-1500 degrees because the temperature will drop when you open the kiln door. Place your piece in the kiln for 2 to 2 1/2 minutes until the lumps are melted. You can peek through the peek hole to see if the lumps are melted. Peeking through the peek hole will allow the kiln to maintain its temperature better.

Get your gloves on and tools ready and open the door. It will be hot. One tool holds the piece down and the other scratches for design. Get out fast, or the cooling glass will stick to your tools. If your tool gets stuck on the enamel shut the door as much as you can on the tools, and allow the temperature to rise, which will allow the enamel to heat up and you can then remove your tools. Remember to shut your door with your gloved hands and not your elbow or hip, because it is hot too. You may need help to do this. The first times you use this technique you may want someone with you to open and close the kiln for you; the process is very fast and hot.

Also, when you remove your tool it may pull some of the enamel off your piece, almost like taffy. This may leave a pulled glass spot on your piece. This is fine, because you need to leave your piece in the kiln for 1-2 more minutes to melt the surface smooth.

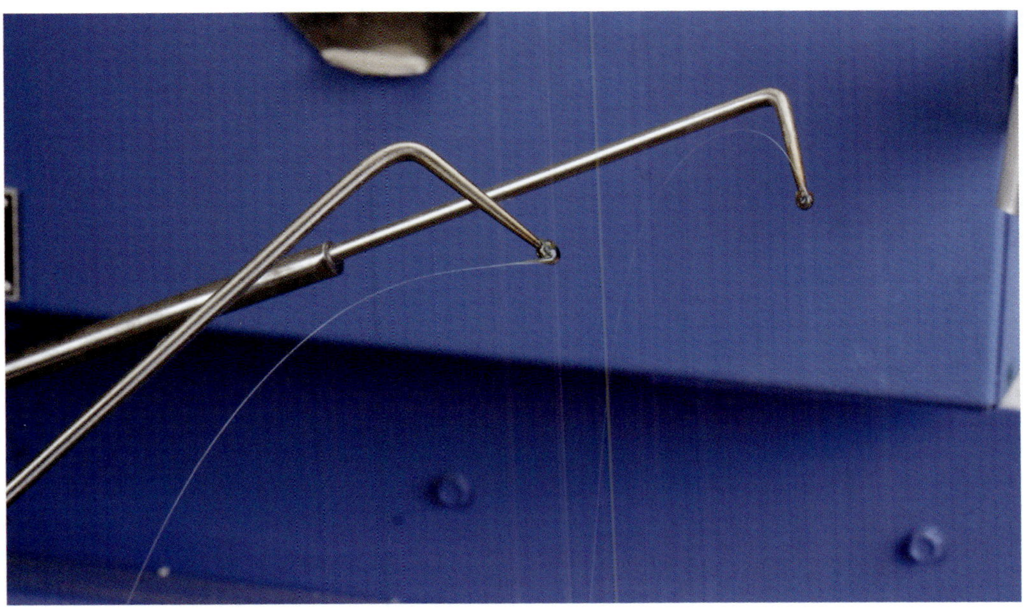

Now remove the piece from the kiln and place it on a hot pad. Wait 5-10 minutes for the true colors to appear. If you don't like the design you may swirl the enamel again. If you want to add colors, you may do that too and swirl again. Just remember, you can't go back, so if you like what you see, don't do anything.

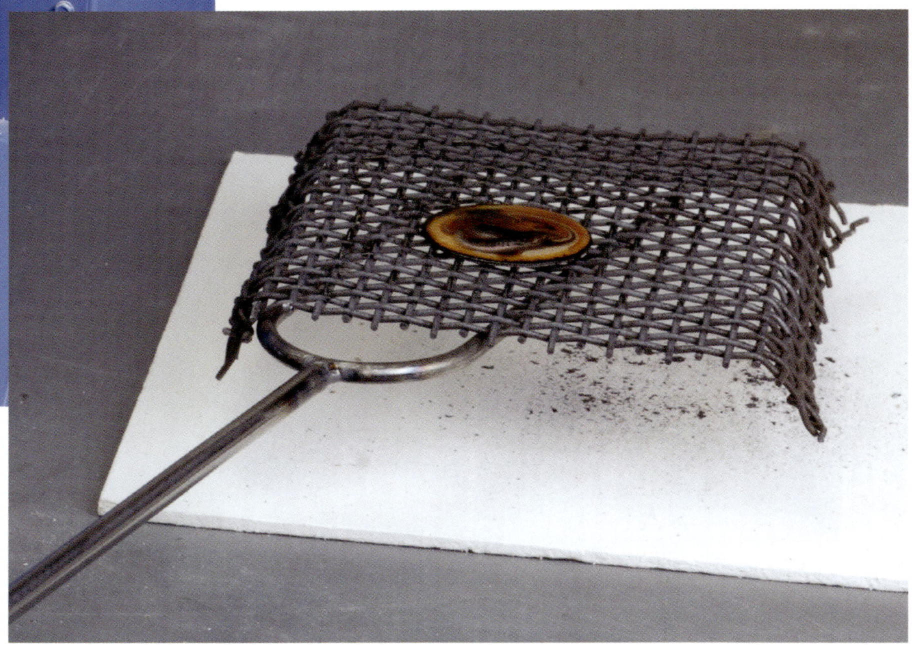

As the piece is cooling, I noticed while I swirled the piece I filled in the hole with enamel. That's why I said you have to be flexible with enamel, as things don't always come out as planned.

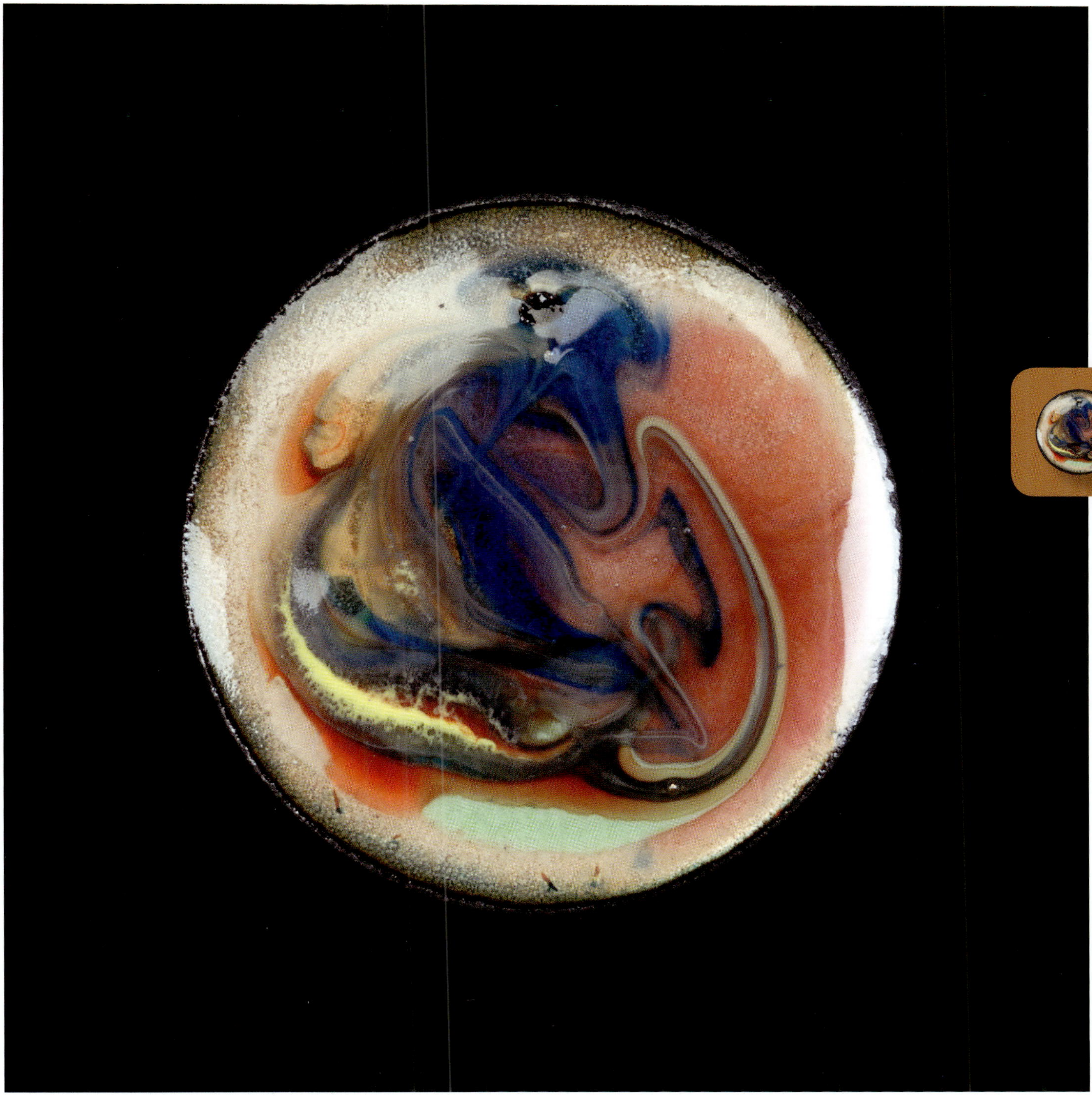

Here is the finished piece.

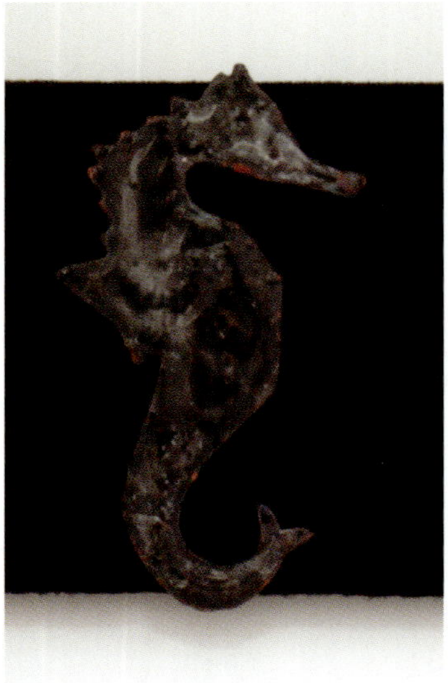

Seahorse Pin

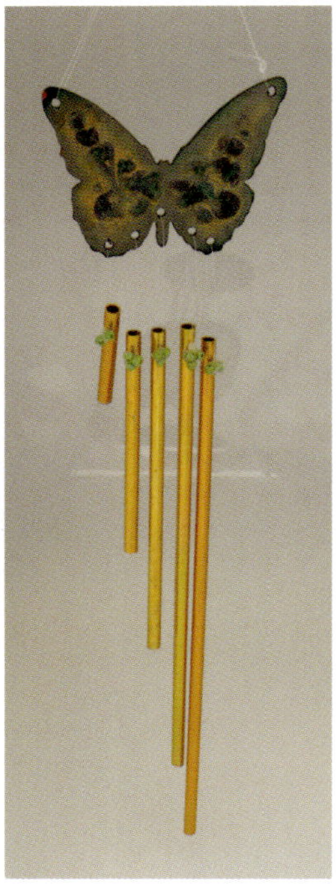

Butterfly Wind Chime

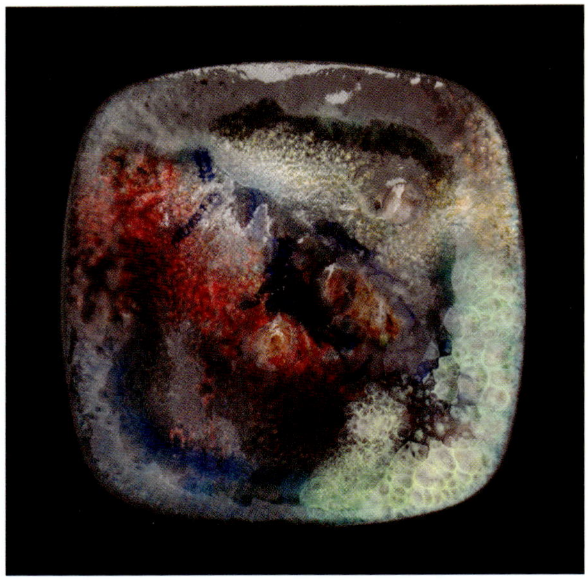

Wall Art

Gallery

"My Universe"

"Blowing out a Candle"

Multi-piece necklace

Necklace with detail

"Three People" Wall Art

Necklace

Framed Wall Art

Wall Art

"My Forest" Wall Art

Multi-piece necklace with beads.

Cloisonné Necklace.

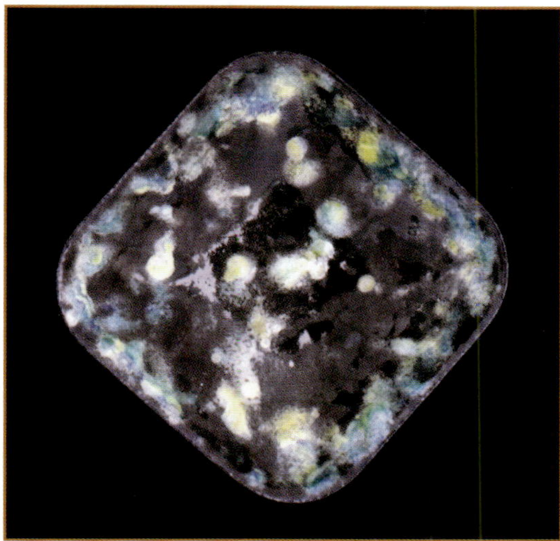

Front of dish

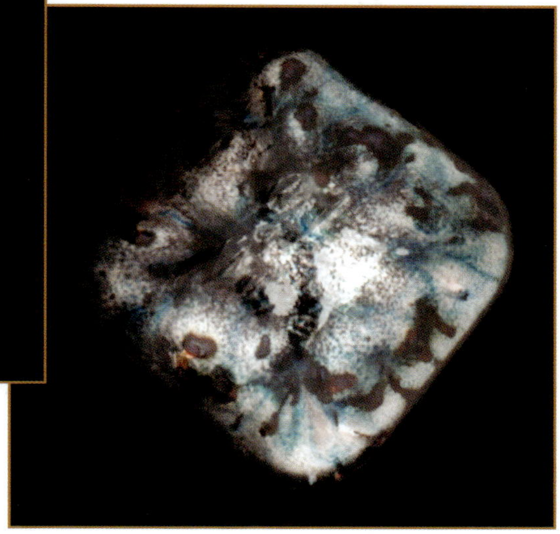

Back of dish

Necklace with layers of copper.

Multi-piece necklace with beads.

Pin and Earring set

"Happy Day"

Enameled ring

Pendant made with enamel crayons.

Reversible butterfly necklace

Bead

Flower pin.